ABOVE: *A modern lead mine, Grove Rake Mine, Rookhope, Northumberland, in 1978.*

COVER: *The Stonedge cupola and its millpond, near Chesterfield, Derbyshire. It is the oldest industrial chimney in the world.*

LEAD AND LEADMINING

Lynn Willies

Shire Publications Ltd

CONTENTS
Lead and its uses 3
Geology and prospecting 7
History of lead mining and smelting 11
Miners, smelters and owners 27

Set in 10 on 9 point Times roman and printed in Great Britain by C. I. Thomas & Sons (Haverfordwest) Ltd, Press Buildings, Merlins Bridge, Haverfordwest.

Copyright © 1982 by Lynn Willies. First published 1982. Shire Album 85. ISBN 0 85263 596 6.
All rights reserved. No part of this publication may be reproduced or transmitted in any form or by any means, electronic or mechanical, including photocopy, recording, or any information storage and retrieval system, without permission in writing from the publishers, Shire Publications Ltd, Cromwell House, Church Street, Princes Risborough, Aylesbury, Bucks, HP17 9AJ, UK.

ACKNOWLEDGEMENTS
Illustrations are acknowledged as follows: Ashover School, page 15; Richard Bird, pages 17 (bottom), 18 (top), 20, 23 (both), 25; Fred Brook, page 27; Paul Deakin, pages 9, 11; Derbyshire Countryside Ltd, page 3; Gwynedd Archives Service, pages 17 (top), 30 (bottom); H. M. Parker, pages 1, 6, 16 (top), 18 (lower two), 22, 24 (top), 26 (bottom), 29 (both), 30 (top); Miss L. M. Simcock and Clwyd Record Office, page 16 (bottom); Trevithick Society and Ironbridge Gorge Museum Trust, page 19; Lynn Willies, front cover and pages 2, 4, 8; from Hunt, *British Mining*, page 24 (bottom). The drawings on pages 10, 12-13, 14 and 21 are by Chris Vince.

A Roman lead ingot, found in 1975 near Ashbourne in Derbyshire. The inscription translates: 'Works of Lutudarum. From Britain: without silver.'

Leadwork at Haddon Hall, Derbyshire, seat of the Manners family, later Dukes of Rutland, who owned the lead rights over a large part of the Peak District.

LEAD AND ITS USES

Lead is a heavy, dull grey metal. Its brightness when freshly cut is usually seen only by those who handle it in their work, and its usefulness is less well known than its unfortunate poisonous properties. Today the 300,000 tons required each year in Britain are mainly imported, but until the late nineteenth century Britain was an important lead producer, and in the eighteenth century it was producing more than any other country — the culmination of almost two thousand years of lead-mining.

Because of its widespread occurrence and the ease of its smelting, lead was one of the very first metals to be discovered and used. It has been found in small pieces in pre-Roman burial mounds, and during the Roman occupation it was commonly used in Britain and exported as ingots or *pigs:* its main uses were for pipes and cisterns, as sheet lead for roofing and baths, as at Aquae Sulis (Bath) and Aquae Arnamentiae (Buxton), and for coffins. Some was converted into pigment: red lead, or minium, and white lead, for paint and even face powders.

With the departure of the Romans, both the need for lead and mining skills went into decline. In medieval times only relatively small quantities were used for roofing churches and some other buildings. A revival came with the increase in the building and rebuilding of great houses in the late sixteenth century. Buildings like Hardwick Hall and Haddon Hall in Derbyshire required vast quantities of lead, on the roof, as gutters and down pipes, for water storage cisterns and pipes, in the great windows, and even on the walls as paint. Increased output from the mines was made possible by new techniques developed in Germany and brought to England by 'Dutchmen' during the reign of

Elizabeth I.

The great age of British mining came in the eighteenth and nineteenth centuries, when Britain was the world's main producer and developed new techniques of mining and smelting. At first much of the output was exported, mainly to the Dutch, who had a world-wide trade in both the metal and red and white lead. After 1780, however, especially when the Napoleonic Wars broke trading links, the manufacture of lead sheet, pipe and shot and of red and white lead grew rapidly both near where the ore was mined and smelted, and at the ports of Bristol, London, Liverpool, Chester, Hull and Newcastle. By 1830 lead ore was being imported, mainly from Spain, and soon cheap piglead began to arrive also. Lead manufacturers found the lower prices highly beneficial but, especially in old mining fields like Derbyshire, mine profits declined to almost nothing, and even cheaper imports from Australia in the early 1880s caused the collapse of the British leadmining industry. Only a few mines managed to continue into the twentieth century. The last of these closed about 1960, and so lead is mined in Britain today only as a by-product of other minerals.

New uses for lead began to develop in the eighteenth and nineteenth centuries, alongside the older ones. The addition of red lead to glass produced a sparkling 'crystal glass', which was more attractive and easier to produce. Lead glazes became common in the pottery industry, for the cheaper stoneware and for products such as Wedgwood's white 'Devonshire', which was dipped in a white lead mixture before firing. It was also used for enamelling metals. Despite the availability of substitutes, more and more lead was used in paints, and as thicker pastes for gaskets on boilers and pipework, and even with canvas to waterproof the roofs of carriages and trams. Horrifyingly, white lead was also sometimes used to adulterate bread to produce a whiter loaf, and lead acetate or sugar of lead to sweeten wine, until legislation was introduced in the 1870s.

During the late nineteenth and twentieth centuries several further uses were discovered and some of the old uses declined as cheaper and more effective substitutes became available. Important new uses of lead were for telegraph and electricity cables and for batteries. As the internal combustion engine developed, lead for batteries became the largest use of all, though because battery lead can easily be recycled tetra-ethyl lead used as an anti-knock agent in petrol has become the greatest net consumer.

An old lead tobacco jar. Lead was also widely used in ornaments.

The manufacture of white lead for paint was perhaps the most dangerous of all lead trades. This woman's job was to build a stack of clay pots containing vinegar and lead. The stack was heated by means of layers of manure and tan bark. Breaking down the stack afterwards, usually done by men, was the most hazardous part.

A vein in Scraithole Mine, Cumbria, up to 4 feet (1.2m) wide, mainly of barite but with a little lead ore.

Methods of searching for lead ore in sixteenth-century Germany, part of the technology introduced to Britain at the time of Elizabeth I. (From Agricola's 'De Re Metallica'.)

GEOLOGY AND PROSPECTING

The main lead ore found is galena (lead sulphide, PbS), though small quantities of other lead minerals have also been mined. It is usually found in veins together with other *gangue* minerals, which were formerly thrown away, especially quartz, calcite, fluorite and barite. Sometimes there is a little silver with the lead, and quantities of zinc and iron ore are also commonly found.

Lead ore is found in all the highland regions of Britain and has been mined in most of them. Modern research has shown that most deposits were formed between 230 and 180 million years ago, that is, in the older rocks which make up the mountains. It is particularly common in limestone, partly because that rock often has suitable cavities in it, but almost any hard rock — sandstone, granite, volcanic — is likely to contain it. As with many metal ores, prospecting is difficult. Important deposits have been found in the Mendips, in the Peak and Lake districts, in the northern Pennines and, with copper and tin deposits, in Cornwall and Devon. Scotland's main area was on the boundary of Dumfries-shire and Lanarkshire, at Wanlockhead and Leadhills. Wales had deposits in all counties, and there were also important deposits on the Isle of Man.

Most of the major veins of lead ore are found in fissures resulting from faults or joints in the rock: they are thus usually near-vertical and can sometimes be traced for several miles, and for a thousand feet (300 m) or more deep. Sometimes the ore is found in pipes or flats, especially in limestone, where the solutions which deposited the ore first dissolved the rock, then replaced it with mineral. Pipes appear

A modern diamond drilling rig, set to bore inclined holes to intersect vertical veins.

like infilled cave passages, which sometimes they are, whilst flats are found as layers between the beds of rock. Both are very common in the Peak District limestone of Derbyshire, but a very large flat was found at Burtree Pasture Mine in the north Pennines replacing the dolerite of the Whin Sill.

The amount of lead ore found varies greatly: it usually forms a thin layer within larger quantities of other minerals but it may *belly out* into a large mass or be *squinted* into an almost non-existent *string*. This made mining of lead ore extremely difficult — quite different from the regularity of coal mining — so it was always referred to as an 'adventure' by the participants. Although the veins are usually vertical, the ore often favours certain beds or horizons — the *bearing beds,* so once these were found great efforts were made to reach them at other mines too.

How the lead ore and other minerals were deposited has puzzled geologists and miners for hundreds of years, and opinion has swung from one view to another many times. Until recently it was thought generally that the minerals resulted from crystallisation from hot-water solutions *rising* from hot granites and other igneous rocks deep under the surface. The modern view accepts this can happen, but most deposits appear to result from hot salty solutions moving *sideways* from deep buried muds, even fifty or hundreds of miles away: these have accumulated in vast thicknesses in the sea and have the property of concentrating heavy metals like lead from the sea. As the mud was compressed to form rock the salty or saline water was squeezed out along faults or fissures, carrying the metals with it, to be deposited in favourable conditions elsewhere.

Prospecting for lead ore and other metals still relies on the same principles today as it did in the past. The divining rod is still used, but its modern equivalent is the measurement of the electrical resistivity of the ground, which will change if a vein is present. Miners and prospectors have always examined the gravel in rivers for traces of lead ore, following them until the source was located, and they studied turned-up soil for signs of ore: today geochemical prospecting can detect minute quantities of lead in stream and soil sediments. Even plants can be indicators of lead, for only a few can survive serious lead poisoning, an example being the vernal sandwort, known as leadwort in mining districts: nowadays geo-botanical prospecting can be done by infra-red photography from satellites. To the geologist, however, the only proof is a sample: nowadays trenching machines and the diamond drill have replaced manual methods of locating veins and the digging of innumerable shafts to try veins at depth. Even with such sophisticated methods however, the actual mining is still a great financial risk. At the Riber Mine near Matlock in the 1950s, for instance, diamond drill holes located a very rich vein, but subsequent mining found this was a small pillar left behind by previous miners to hold back water.

The marks of miners' picks, dating from about 1745, in Ringing Rake Sough, Matlock, Derbyshire. Work advanced about 2 inches (50 mm) a day.

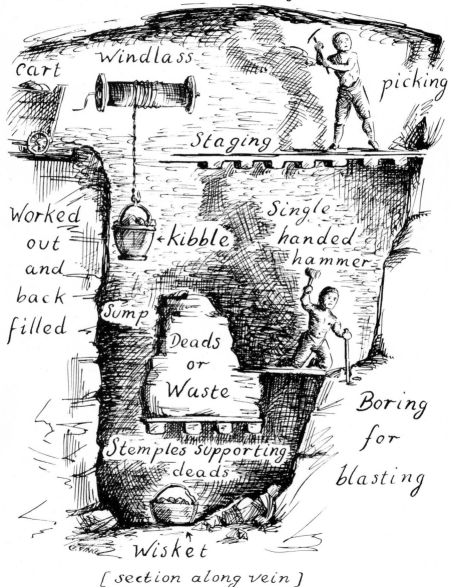

Miners' tools, abandoned about 1850 in Gentlewoman's Mine, Matlock, Derbyshire.

HISTORY OF LEAD MINING AND SMELTING

Evidence of Roman mining comes from the thirty or so lead ingots that have been found and the smelting hearths that have been located, three in Derbyshire, and one at Pentre in North Wales. Roman mining has been rather dubiously claimed as the origin of the lines of shallow pits seen in the Mendips, in North Wales and in Derbyshire, and even more unlikely, of caverns underground. Most workings, even if fairly recent, have at least three phases of activity, as later techniques made it economic for material left behind to be worked again. Any easily accessible lead-bearing ground was turned over again and again in the nineteenth century, to find low-grade ore to feed an improved furnace known as the Spanish slag hearth.

Probably the earliest workings which can satisfactorily be dated belong to the seventeenth century, when levels began to be driven through rock that did not bear ore in order to drain flooded mines — there was no incentive to destroy these later. These levels were generally driven using hand tools only — the pick and wedge and hammer – occasionally assisted, if the rock was dry, by *fire setting* (lighting a fire against the rock to cause it to become more friable). These early tunnels were very small, especially if fire-set, being no more than a foot (300 mm) wide, and 3 feet (900 mm) high, often less: this type of work continued into the mid eighteenth century and even later, and its most developed form was the *coffin level,* which is just large enough

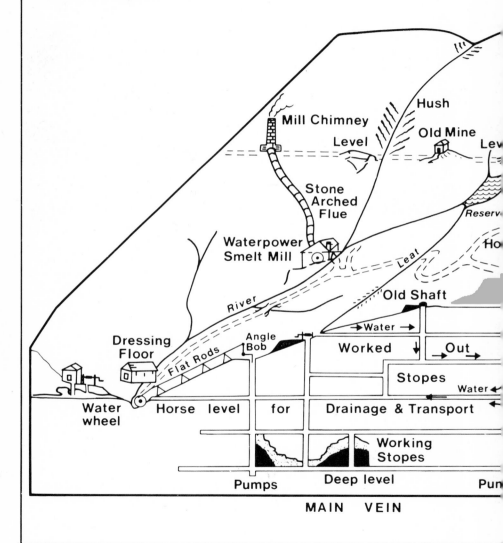

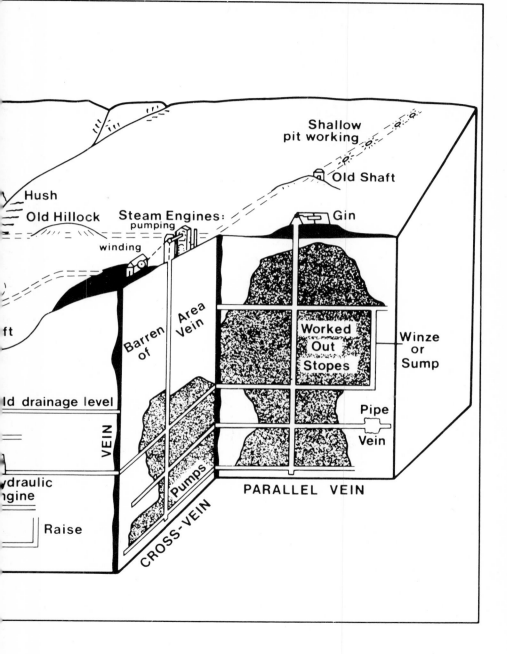

Methods of winding.

for a man to walk through. Until the black powder (gunpowder) and other new mining techniques were brought into common use in the eighteenth century, the methods of mining had remained almost unchanged for centuries: indeed evidence of Roman lead-mining in Spain suggests that the Romans were more advanced than the miners of much more recent periods.

When a vein was discovered, it was first followed down from the surface by trenching or shallow pitting; any ore was removed, and the waste thrown up and aside. When the pit or trench became too deep, short timbers or *stemples* were put across the vein between the solid walls to form a platform and the waste was piled on them, and mining proper started underneath: this is *stoping* — a method of working a vein in steps, stacking waste or *deads* behind and overhead, leaving *gates* (passages) and *sumps* or *winzes* (shafts not open to the 'day' or 'grass') for access and ventilation. These pits were connected to the surface by shafts equipped with a jack-roll or windlass (*stow* in Derbyshire), to wind out *kibbles* or buckets of ore and waste. As the mine was predominantly vertical, two openings were usually sufficient to provide ventilation, though often the air was extremely bad, and the candle sometimes had to be placed some distance from the miner so that it would burn at all!

Where there was water it might be wound out by hand, or if the mine was close to a valley it might be drained by a short level or *sough*. Only a very few mines used a horse gin to wind water or operate a primitive pump, and even fewer had waterwheel-powered pumps until the eighteenth century. Where water could be accumulated, then sometimes *hushing* was used: water stored in a dam was released to pour down a hillside along a vein — tearing out the weathered rock to expose the solid rock and vein, and by a process of natural separation allowing lumps of heavy galena to be sorted lower downhill from the lighter waste stuff. The process was repeated again and again. Many such hushes are found in the mid Pennines around Reeth and Arkengarthdale, and further north near Alston.

Most of the easily found veins had been

located by the eighteenth century, and though such simple methods continued in use until the twentieth century, more and more hopelessly, the problem of water at increased depths required ever more capital to deal with it. The most important means of solving the problem was the steam engine. Captain Savery developed a simple fire engine about 1695, but this was only occasionally used – as near Cromford in Derbyshire about 1710. However, Thomas Newcomen's atmospheric engine provided a useful power source from 1712 onwards: the first on a lead mine was at Winster in Derbyshire in 1717, and by 1730 the Yatestoop mine there had three engines. With an annual production of almost 3,000 tons at its peak this was briefly the most important mine in Britain. Long drainage soughs – a mile (1.6 km) or more in length – were developed in Derbyshire and in the first half of the eighteenth century these skills were exported into Wales, Yorkshire and the north Pennines.

At about the same time the London Lead Company, 'The Governor and Company for smelting down lead with pitt-coal or sea coal', was developing both smelting

A Newcomen engine first erected about 1748 by the London Lead Company at Millclose Mine near Matlock and later moved to Gregory Mine, Ashover, Derbyshire.

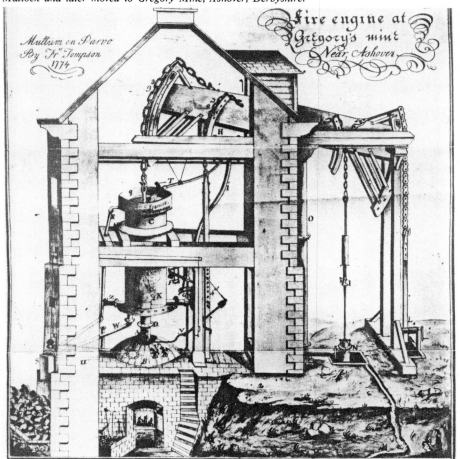

A Roman smelting hearth from near Duffield, Derbyshire.

and mining in North Wales and spreading its operations into Derbyshire and, more importantly in the long term, into the north Pennines. The smelting of lead had gone through several stages of development: the Romans seem to have used small hearths, rather like a crude smith's hearth with the blast provided by foot bellows. From the post-Roman period to the sixteenth century, the blast was provided by the wind on the brows of high hills, sites known as *boles* or *baile-hills* in Yorkshire. In the sixteenth century, as a result of German influence, water power was adopted for the blast in the ore hearth, a practice which continued in some areas until the twentieth century. All these methods relied on wood, dried in a kiln, for fuel, together with a certain amount of peat if available. The London Lead Company developed a coal-fired

The smelting works at Gadlis, North Wales, about 1720. It was built by the London Lead Company and was the site of the first successful use of the coal-fired reverberatory furnace or cupola.

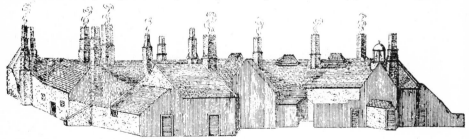

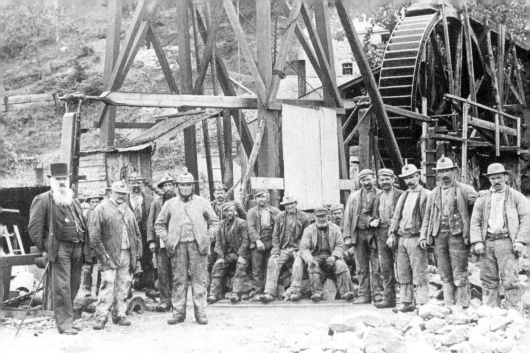

ABOVE: *Miners at Trecastell Mine, Conwy, North Wales. Note the tallow dips or candles attached with a blob of clay to their hard hats. The wheel is linked to the pumps in the shaft by the usual flat-rods.*
BELOW: *Lock gates in Magpie Sough, Derbyshire, allowed the use of boats for transport. This sough was driven in about 1873-81, and dynamite was used for this section.*

ABOVE: *Massive stone arching in Hillcarr Sough, Derbyshire, built in 1766-83. The sough is over 4 miles (6.5 km) long.*
BELOW LEFT: *Stone arching in a horse level at Smallcleugh Mine, Cumbria. Many miles of such levels were driven by the London Lead Company and other companies in the North Pennines.*
BELOW RIGHT: *A timbered horse level in East End Allenheads Mine, Cumbria. Timber has a shorter life than stone but was cheaper.*

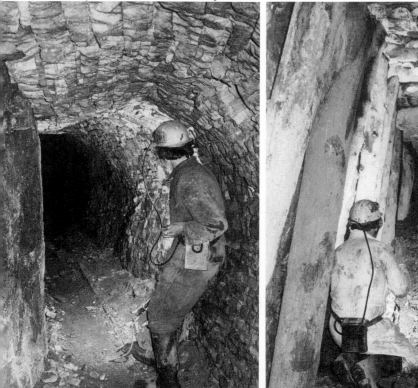

reverberatory furnace – more efficient, not needing water power, but requiring a much larger throughput, which encouraged it to begin large-scale mining to ensure its ore supply. It had the capital to introduce the new mining technology, and though it failed in the much exploited Derbyshire field, its long-term development of the relatively unexploited northern Pennines around Alston and Nenthead was outstandingly successful. Its most important contribution to mining was the systematic development of horse levels, which located new veins, drained them to great depths and made underground haulage to a central ore dressing and smelting site profitable.

Though the steam engine was probably the most important new method available its very expensive first cost and its high running costs encouraged the use of alternatives, especially long drainage levels. In Derbyshire the Hillcarr Sough system extends about $4\frac{1}{2}$ miles (7 km), whilst at Nenthead a system of levels extended over 20 miles (32 km) by 1820. Whenever water power was available it too was used, but though it was cheap it was unfortunately at its lowest in summer when the mines were most easily pumped: Wales, the mid and north Pennines and the Lake District have both high rainfall and suitable geology and topography for waterwheels, and dams, leats and wheelpits are still common in the mining areas. Waterwheels could not always be placed convenient to the mines but could be connected by *flat-rods* running on rollers to the shaft, where the horizontal motion was transformed to vertical by an angle-bob. In many cases waterwheels were placed underground, with water provided from a natural flow or down a convenient shaft and leaving the mine through a sough or drainage level. In a few cases a device known as a *balance-tub* or *balance-bucket engine* was used, the necessary pumping motion being obtained from a beam pivoted at the centre, with an alternately filled and emptied tub or bucket at one end. Whatever the source of power, the motion was transmitted via vertical rods in the shaft to the iron or wooden pumps below, pumping to the surface or to the drainage level.

Large nineteenth-century mines were characterised by a new regularity: despite the use of black powder and the introduction of carts and then wagons running

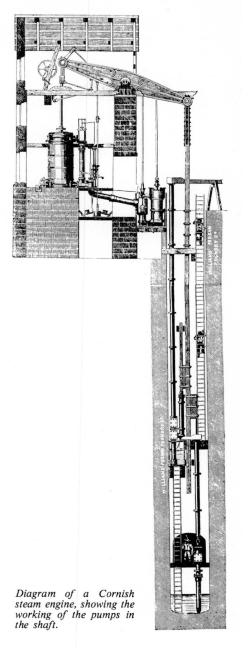

Diagram of a Cornish steam engine, showing the working of the pumps in the shaft.

The Cornish engine house at Balters Shaft, West Chiverton Mine, one of Cornwall's most successful lead mines.

on rails and even boats on underground canals, most mines until then were rather like dozens of small mines put together, with teams of miners working independently of each other, their efforts united only by a common drainage system. Usually only the best ore was taken out, much of the sorting being done in the mine itself. Change came from a number of developments. Steam replaced horse power for winding and could compete with and complement the large wagons used in horse levels. New washing (*dressing*) methods and improvements in smelting, especially the new blast hearths, could treat ore of poorer quality. It was found that with better haulage larger tunnels became economic, and it was more efficient to have two or even three men combining to drill shotholes, using heavy (two-handed) hammers, rather than a single man using a light mallet. With better haulage and improved ore dressing, instead of *underhand stoping* — working down from the top — *overhand* methods could be used, in which the vein could be drilled and blasted down from below, using gravity to load the broken stuff directly into wagons. Sorting of ore was done mechanically or by cheap labour at the surface, rather than underground. Pumping methods, however, remained the key improvement, for as mines got deeper they were generally below valley level and the drainage adits of the previous century.

Lead mining was dominated in the North of England by the London Lead Company, with a few other large concerns like the Blackett-Beaumonts in Weardale and Allendale and the Old Gang Company and the AD Company in the mid Pennines. In Derbyshire, the South-west, Wales and Grassington, John Taylor and his Cornish methods were predominant.

Taylor advocated systematic highly capitalised mining: he was responsible for little that was entirely new but used developments like the improved Cornish pumping engine as much as possible.

By the mid nineteenth century, by which time first canals and then railways had lowered fuel costs, the Cornish engine was twenty or more times as efficient as the early Newcomen engines, as a result of the work of the Cornish engineers Trevithick, Woolf and Grose. Large heads of water, made possible by extensive drainage systems penetrating deep below upland rivers, were used by water-pressure engines placed underground — Derbyshire had six, including the Guy Engine (the largest ever made), operating on the Alport Mines, and they were common in North Wales and the northern Pennines. Massive waterwheels, some 50 feet (15 m) or more in diameter (one can still be seen at Laxey on the Isle of Man), made the most of surface water. At Alport in Derbyshire and in Wales at the Minerva Mines, it was possible to pump water at rates of 5,000 to 6,000 gallons (20,000 to 30,000 litres) a minute — equivalent to what previously had been tapped by soughs. Mines like these required large outputs to sustain the costs of such pumping and the associated large labour forces, and the lives of many mines, despite large investments, were short.

At the surface changes were just as apparent as underground. To the end of the

In its underground chamber, this hydraulic engine for pumping in Sir Francis Level, Swaledale, North Yorkshire, was installed in 1881.

eighteenth century small mines requiring little more than a small hut, or *coe,* and a windlass were still the commonest, and even large mines, unless there was a steam engine, could make do with a horse gin and a few simple buildings to store ore and materials. Dressing processes were simple, comprising hand sorting and crushing, with separation of ore from lighter waste by plunging in a sieve in water, and buddling of fine-sized stuff as a slurry in a stone-lined trench. This work was mainly done by women or old men, assisted by boys.

In the nineteenth century the use of steam or water-powered winding and pumping engines became normal, and the number of small mines dropped drastically, since the economics of mining made it difficult to earn a living, and large mines bought out or otherwise cleared small miners off their titles, often assisted by local mineral owners, who naturally favoured the likeliest provider of high royalties. Larger quantities of material brought out encouraged mechanisation of dressing processes, and mechanisation encouraged centralisation on to large dressing floors.

Formerly the ore had been crushed with spalling hammers if large, and then by bucking hammers down to pea-size. Then horse crushers — a heavy edge-wheel pulled around a stone or iron paved circle — were used, then powered crushing rollers, and after 1870 Blake-Marsden type jaw crushers, introduced from America.

The heavy work of plunging an ore-filled sieve in water to separate heavy ore and light waste into layers was replaced first by the pole sieve, in which the weight was taken by a counterbalanced pole, then by the Cornish jig, which allowed use of steam or water power. Buddling, in which a slurry of fine ore in water was run down a slope to separate heavy from light material, was changed more slowly, but it developed through Cornish round buddles, improved and extended trunk-buddling systems, to the eventual use of shaking tables. Ironically, this work, which when heavy was done predominantly by women, was done mainly by men when the labour was eased and instead of being outside the washing floors took place inside a 'mill'.

Thus by the later nineteenth century the larger mine had substantial waterwheels, with perhaps steam winding and pumping engines as well or instead, each in its own house; there would be stores for mining materials, powder and for the ore, with buildings over the washing floors, all connected by narrow rail systems to the shaft and waste heaps. Nearby were the mine offices, the agent's house and, if the mine was remote, a barrack-like *lodging shop* where men stayed during the week. In many cases a village grew up around the mine, and not far away there would be a smelter, with its furnaces and associated chimneys and possibly a long flue to carry the poisonous fumes a mile or so to a high point where their discharge would be less harmful.

Leadmining and many leadmining settlements all but died out in the 1880s, and only a few companies kept going: in Derbyshire the sole survivor of the thousands of mines was Millclose Mine, with a recently discovered rich vein. In

The Lady Isabella waterwheel, $72\frac{1}{2}$ feet (22 m) in diameter, was erected in 1854 at the Laxey Mine, Isle of Man. The angle-bob at the shaft (right) changes the horizontal movement of the connecting flat-rods to vertical.

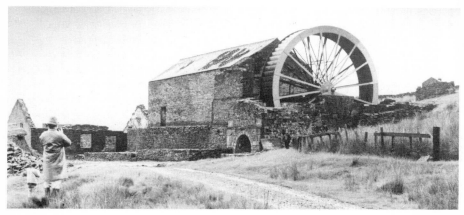

ABOVE: *The Killhope Wheel, Northumberland, was used to operate crushing rollers.*

LEFT: *Blake's stone crusher, introduced about 1870 (from an advertisement in the Mining Journal of 18th June 1870). Note the children working at the picking table.*

BELOW: *Sorting ore in a well organised mill about 1860. The ore is first swilled clean (right), then sized by a revolving sieve or trommel (centre), and finally hand-picked on a rotating picking table (left).*

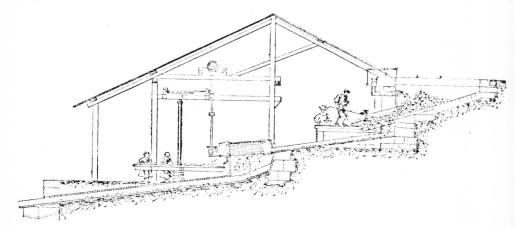

A Pelton wheel, a water-powered turbine, at Greenside Mine, Cumbria, about 1920.

North Wales the Halkyn mines through several reorganisations continued their policy of deep tunnel drainage. In the Lake District Greenside Mine near Helvellyn installed the first hydro-electric plant in mining, and in the Alston area the Weardale Lead Company took over from the Blackett-Beaumonts, and the London Lead interests were given up first to the Tynehead and Nenthead Zinc Company, and then to the Vieille Montagne Company of Belgium: both companies combined zinc with lead mining. In Scotland the Wanlockhead Lead Mining Company was revived in 1906. Otherwise lead production was negligible, the almost accidental by-product of other mining or quarrying operations. All these mines had a history, or at least the hope, of sufficient ore to repay high costs at times of low prices; they had modern equipment and shafts, and no doubt tenacious owners, capable of extracting and either using or selling subsidiary minerals as well as lead, especially zinc, which was in considerable demand.

In the late nineteenth century compressed air drills and Nobel's dynamite reduced the costs of tunnelling and sinking: the Milwr Sea Level Tunnel achieved the world record in 1930 with 2,037 feet (611 m) in fourteen weeks. Electricity and, to a lesser extent, compressed air gave new flexibility to winding and pumping underground — Millclose for example was eventually able to work 3 miles (4.8 km) from its shaft. Both could often be produced in remote areas by water power, using Pelton wheels or other turbines, or, as at Millclose, first by the use of diesel engines and then from electricity supply companies. The most successful mines were Millclose, Halkyn and Greenside, producing in the 1930s the highest ever British output of ore and at prices that had not been lower since about 1700. Millclose, Britain's largest ever mine, suffered disastrous flooding in 1937 and had effectively closed by 1939; Greenside finally exhausted its reserves in 1962, and Halkyn finished commercial mining about the same time, though there has been some production since, associated with maintenance of the Milwr Tunnel for water supply. At Millclose, however, a lead smelter built just before the mine's closure now recovers and resmelts battery lead.

Since the Second World War leadmining has been at an even lower ebb, despite several attempted revivals using modern (diamond drill) prospecting methods. In North Wales Parc Mine operated on a large and over-optimistic scale in the 1950s. In Derbyshire attempts to reopen the Riber Mine and the Magpie Mine were soon abandoned. However, lead is being successfully mined in the north Pennines and Derbyshire, as a by-product of the mining and quarrying of fluorspar, barite and calcite. It is unlikely to be mined exclusively again unless it becomes a very scarce material.

ABOVE: *Millclose Mine near Matlock, Derbyshire, in the 1920s. The mine became Britain's largest ever lead mine, with a huge output in the 1930s.*

BELOW: *Modern mining. The Allenheads Mine, Northumberland, worked for fluorspar by the British Steel Corporation.*

The twentieth-century remains of a once flourishing industry: desolate tips and a ruined engine house at Snailbeach Mine, Shropshire.

MINERS, SMELTERS AND OWNERS

Little is known of the people who mined lead in Roman times, though their settlements and road systems are observable from aerial survey of the mining areas, for example Charterhouse in Somerset. The Romans may have taken direct possession of the mines, working them with slaves, as is sometimes claimed, or, more likely, they controlled smelting and used a royalty or tax system, encouraging the private enterprise of individuals to undertake the risk of mining. With a few exceptions this was the system used by the Crown or landowners during and after the middle ages.

In the Mendips and Derbyshire for instance, medieval laws probably of Saxon origin governed the right to mine, allowing individuals to claim *meers* of ground along veins, and in return for this and other privileges the miners were charged various duties or royalties (such as lot and cope in Derbyshire) based on the output. Areas in North Wales and Yorkshire followed their example, but some landowners preferred to lease their rights for fixed periods to either individuals or companies. However, even where this was so, the mining was normally carried out as part of a 'bargain' by the working miner, rather than as a paid employee. Generally therefore the miner was notably independent, and virtually self-employed until the time of the large companies in the nineteenth century. Various strikes at that time, in Derbyshire, North Wales and Yorkshire, nearly all concerned the reduction of this independence by the companies forcing miners to work a regular eight-hour day instead of the six hours worked previously, though double-shifting had been common. The miner was thus able to combine mining with a smallholding, and the many small fields still to be seen close to old mining settlements are evidence of this.

Originally the mining laws governed almost the whole of the miners' work; they were a system of rough justice in what were then still wild areas. In the Mendips, for instance, the thief of lead ore for the third offence was locked in his coe or storehouse, and the thatch was set alight. In Derbyshire his hand was fixed by a knife to the stow or windlass, and there he died or cut his hand free with another knife. Only Derbyshire still has its customs, though the more peculiar were abolished in 1851 and 1852, but as with other areas these became modified to permit large companies to work, and the life of the miner typically became bound to the company and its agents, though it was long common for him to work a small mine on his own account in his spare time.

Generally the large mines of the eighteenth and nineteenth centuries had a bargain or setting day every six or eight weeks, at which teams of miners, known as *pares* or *copers* in different areas, bid against each other for various tasks on the mine. The mine agent occasionally came to a private agreement with a team but usually would accept the lowest bid: this was either a *cope* or *tribute* bargain, where payment was according to the ore produced, or a *driving* or *tut* bargain based on the amount of rock or vein extracted. In both cases the provision of tools and candles and the winding and dressing of the ore was left to the miners, so that they bore a great deal of the risk. Sometimes the miners came across unexpectedly rich ore, and in North Wales in about 1850 six men earned the enormous sum of £120 in a month, when others made only £1 to £2 each in the same period. In mining 'adventures' such a gain was more than balanced by hundreds of poor bargains.

Despite the feeling of adventure, the miners' work was usually both hard and monotonous. To get to and from work the miner might have to climb on ladders or stemples for hundreds of feet. In tough rock the use of pick and wedge advanced a tunnel only an inch or two a day, and though it was faster using black powder the hours of 'beating the borer' in often cramped positions were extremely hard. Damp and muddy, often cold and wet conditions led to rheumatism, whilst bad smoke-filled air brought on 'winding' or bad chest ailments. In the sandstones of North Wales and the mid and north Pennines, silicosis made men prematurely old at thirty, and many died before they were forty. The problem was worsened when compressed air drills were introduced. Accidents caused by roof falls or explosions of firedamp were much less frequent than in coal mines, and surprisingly lead poisoning, at least in a severe form, was not common amongst miners. In most areas the miners lived fairly close to the mines, the men being drawn from the partly agricultural population, and migrant miners finding lodgings, and frequently wives, from their families. Agriculture tied men to local mines, and whereas younger men considered travelling normal, the inevitable failure or decline of a local mine in many cases caused extreme hardship as families struggled to maintain a living without moving. Some mines which were remote, as in the northern Pennines, had lodging shops for men to stay during the week and the atmosphere inside some of these was claimed to be worse than in the mine itself.

The work at the surface was also hard. Until the nineteenth century the dressing of ore and winding was normally done by women and children. It involved bending and lifting heavy weights, in wet and often cold conditions, in upland areas with no proper shelter. Even when the processes were mechanised and carried out in buildings, these were normally partly open to the wind, whilst the hours worked, twelve or thirteen a day, were much longer than for the miners.

Some miners had jobs of higher status. The mine agent was, on a large mine, a much respected figure, and though his salary might not be higher than the miners', he often managed more than one mine. His income was all too frequently increased by dubious means, the subject of many complaints — supplying goods to the mine at high prices, favouring bids from members of his family, or even insisting miners' purchases were to come from his shop. Most agents came up from the ranks of the miners but a few, after a short apprenticeship, were appointed by their family who owned or controlled the mine. In Derbyshire about 1825 William Wyatt was appointed agent of several mines in this way by the age of twenty-one, becoming one of the principal and best known agents in the area. Every mine he managed made considerable losses, though so did most others!

ABOVE: *Rotherhope Fell miners in the late nineteenth century. The group probably represents a team or 'pare' of miners, perhaps with the mine agent.*
BELOW: *Miners at Smallcleugh, Cumbria, in 1897, when the mine was operated by the Vieille Montagne Company.*

ABOVE: *The lodging shop at Nentsbury Haggs Mine, Cumbria.*

BELOW: *Miners' barracks or lodging shop at Frongoch Mine, mid Wales, about 1900. These were clean and well ordered.*

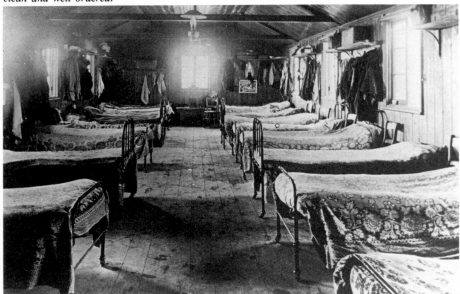

As mines became larger and more mechanised, engineers and enginemen were required; they had high status but worked long hours. Very large mines run on the Cornish system had captains, or supervisors, in charge of each department: underground, dressing, surveying, and so on.

Smelters had little to do with individual mines, except the larger ones, and so often formed separate communities. Their furnaces had to be kept going twenty-four hours a day, and so though their work was less heavy than that of the miners they usually had to work long hours. Though much depended on the design of the mill and the ventilation, smelting was often a dangerous occupation, subject to severe lead poisoning. The bellows of the ore hearth, for instance, frequently blew fumes directly into the smelters' faces. Where red and white lead pigments were made the problems were even worse: at one Sheffield works in the 1830s, according to the doctor, 'men died like sheep'.

Until the late period miners were often also the mine owners, though it was rare for them to have more than modest prosperity. In the eighteenth and nineteenth centuries, however, much capital was needed to work large mines and so ownership became separate from actual mining. Often this gave considerable power to men like Wyatt, who could control the mine, viewing shareholders, nominally the owners, as fit only to supply more money – though this had to be done with finesse. A great deal of the capital came from local shareholders, though the greatest and longest lasting company of all, the London Lead Company, drew its capital originally from London Quakers. In the nineteenth century the practice of 'floating' mining companies became widespread, especially in Wales and the South-west, to the unfortunate extent that someone in disagreeable circumstances was said to have about as much chance as a 'Londoner in Cornwall'. Often mines were floated by smelters, desiring to increase the ore mined without incurring too much of the risk themselves, and only in the mid and north Pennines did the same companies combine the risks of both mining and smelting.

Miners and owners were therefore often unknown to each other, and at times of distress this might make the hardship more severe. Frequently, however, the problems were at least lessened: the London Lead Company was famed for its philanthropic activities in the north Pennines, buying and stabilising the price of grain when it was becoming expensive, encouraging and contributing to the sick clubs which developed in most mining communities, and providing schools for employees' children. Landowners, remembering both their traditions of nobility and no doubt future mining royalties, frequently and generously contributed to emergency arrangements during distress or after accidents. During the terrible years around 1830 relief measures such as soup kitchens were established in most leadmining areas: the owners were distressed by the condition of the miners and their families and feared the loss of accumulated skills; the men clung to their smallholdings, avoiding to the last moment the obvious solution to their problem – a move to coalmining.

WARNING

Mining areas are often dangerous, with open shafts and ground liable to sudden subsidence. The greatest care is required, and normally the visitor should keep to the footpath. For the inexperienced, entering old mines is extremely dangerous – they depend all too frequently on rotting timber for support. Keep out unless in the company of experienced explorers, such as members of a mining society.

MINING SOCIETIES

Most mining areas have one local society or more. For local information write (enclosing a stamped addressed envelope) to:

National Association of Mining History Organisations. Secretary: M. Gill, 38 Main Street, Sutton, Keighley, West Yorkshire.

The two principal societies, both of which publish journals, newsletters and special publications, as well as undertaking exploration and conservation, are:

Peak District Mines Historical Society Ltd, c/o Peak District Mining Museum, Matlock Bath, Derbyshire.

Northern Mines Research Society, c/o Mr J. H. McNeil, Ythan Bank, 12 Woodville Road, Brierfield, Nelson, Lancashire.

PLACES TO VISIT

Craven Museum, Town Hall, High Street, Skipton, North Yorkshire. Telephone: Skipton (0756) 4079.

Earby Mines Museum, Old Grammar School, School Lane, Earby, Colne, Lancashire. Telephone: Earby (028 284) 3210.

Llynwernog Silver Lead Mine, Ponterwyd, Aberystwyth, Dyfed. Telephone: Ponterwyd (097 085) 620.

Peak District Mining Museum, The Pavilion, Matlock Bath, Derbyshire. Telephone: Matlock (0629) 3834.

Richmondshire Museum, Ryders Wynd, Richmond, North Yorkshire.

Wanlockhead Museum Trust, Goldscaur Row, Wanlockhead, Dumfries-shire. Telephone: Leadhills (065 94) 387.

Visits to surface remains and underground workings are possible in most areas: see the books listed below or call at one of the above museums for access details.

FURTHER READING

The best national surveys of mining remains in photographs are contained in Richard Bird *Yesterday's Golcondas* (Moorland, 1977) and *Britain's Old Metal Mines* (Bradford Barton, 1974). Introductory works, for the Peak District, include Ford and Rieuwerts *Lead Mining in the Peak District* (Peak Park Planning Board, 1975), Parker and Willies *Peakland Lead Mines and Miners* (Moorland, 1979), R. Flindall and A. Hayes *Caverns and Mines of Matlock Bath* (Moorland, 1976), N. Kirkham *Derbyshire Lead Mining through the Centuries* (Bradford Barton, 1968), and Willies, Roche, Worley and Ford *The History of Magpie Mine* (Peak District Mines Historical Society, 1980). The Peak District Mines Historical Society *Bulletin* has many articles on the area. For the North of England Raistrick and Jennings *Lead Mining in the Pennines* (Longmans, 1965) is complemented by Raistrick *Lead Mining in the Mid-Pennines* (Bradford Barton, 1973) and his *Lead Industry of Wensleydale and Swaledale* (two volumes, Moorland, 1976), and C. J. Hunt *Lead Miners of the Northern Pennines*. R. Clough *Smelting Mills of the Yorkshire Dales and Northern Pennines* has detailed architectural sketches of many smelting remains (Keighley, 1980). The *Memoirs and Transactions* of the Northern Mine Research Society contain much information on this and other areas, especially Wales. Mid Wales has been reasonably covered by David Bick *Old Metal Mines of Mid-Wales* (Poundhouse, Newent, 1974), North Wales in C. J. Williams *Metal Mines of North Wales* (Charter Press, Rhuddlan, Clwyd, 1980), and South Wales in G. W. Hall *Metal Mines of Southern Wales* (Westbury on Severn, 1971). Fred Brook and Martin Allbutt *The Shropshire Lead Mines* (Moorland, 1973) covers that county. Publications on the South-west generally refer only slightly to lead, concentrating on tin and copper: however D. B. Barton *Cornish Beam Engine* (1969) has wider implications, and his other works, published by Bradford Barton, Truro, have much related material. For other areas Bird's two books form the major starting point.